595.7   Graham, Ada
GRA
        Busy bugs

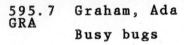   7706

DATE			
	TUA		

MORENO VALLEY UNIFIED SCHOOL
13911 PERRIS BLVD
MORENO VALLEY CA
923880000

09/23/89

                817425   13178

© THE BAKER & TAYLOR CO.

BUSY BUGS

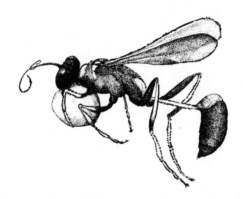

Ada & Frank Graham

Illustrated by D. D. Tyler

DODD, MEAD & COMPANY   NEW YORK

Text copyright © 1983 by Ada and Frank Graham
Illustrations copyright © 1983 by D. D. Tyler
All rights reserved · No part of this book may be reproduced
in any form without permission in writing from the publisher
Printed in the United States of America
2 3 4 5 6 7 8 9 10

Library of Congress Cataloging in Publication Data
Graham, Ada.
　Busy bugs.
　Summary: The busiest life periods and special
body parts of fourteen kinds of insects, ranging from
field crickets to tent caterpillars, are examined.
　1. Insects — Juvenile literature. [1. Insects —
Habits and behavior] I. Graham, Frank, 1925-
II. Tyler, D. D., ill. III. Title.
QL467.2.G7　1983　595.7　82-22085
ISBN 0-396-08126-6

CONTENTS

Busy Bugs	5	Honey Bee	34
Mud Dauber Wasp	6	Tent Caterpillar	40
Field Cricket	10	House Fly	44
Burying Beetle	14	Leaf Miner	48
Dragonfly	18	Monarch Butterfly	52
Aphids	22	Acorn Weevil	56
Firefly	26	Caddisfly	60
Honeypot Ant	30	Scientific Names	64

*To my best love, Hank, and my most reliable fans,
Grandma Mac, Phil, Zach, and Kate* — D.D.T.

BUSY BUGS

All of the creatures in this book are insects. None are true bugs, scientifically speaking. But *bug* is a word that many people use when talking about any insect. And all insects are certainly busy. They are busy eating and growing, or finding mates and providing food and shelter for their young.

We have chosen fourteen kinds of insects to show just a few of the ways in which these remarkable creatures lead successful lives. We look at the period in their lives when they are busiest, and explain how they make use of special parts of their bodies to get their jobs done.

The mud dauber wasp uses strong front legs and mandibles, or jaws, to build cells for its young. The field cricket listens to mating calls with its knees. The tent caterpillar uses a special tool below its mouth, called a spinneret, to lay down silk.

There are more than a million kinds of insects in the world. At first they seem strange and confusing to us. But when we watch them closely, we see how they use their legs and jaws and antennae. We see them hunting food, or looking for a mate, or finding a good place to lay eggs.

Next time you see an insect, remember that it is not wandering around without something to do. All insects are very busy bugs.

MUD DAUBER WASP

The mud dauber wasp goes to work by standing on her head. She has found what she was looking for—a muddy place on the ground.

Her long hind legs tip her body far forward. With her strong mandibles, or jaws, and her front legs she rolls lumps of mud into balls the size of a pea. The wasp is getting ready to build a cell in which to lay her egg.

The muddy ground shows how busy the wasp has been. She leaves footprints wherever she lands. She makes tiny holes in the ground where she removes the mud.

The wasp lifts a mud ball and flies away with it. She holds it with her mandibles and front legs. Powerful muscles in her thorax, the middle section of her body, control the movement of her legs and the beating of her wings.

The mud dauber usually attaches her cell to the side of a building. She flies to the wall, bringing mud which she daubs, or smears, on it. The tiny mud balls she brings later on stick to this wet coating.

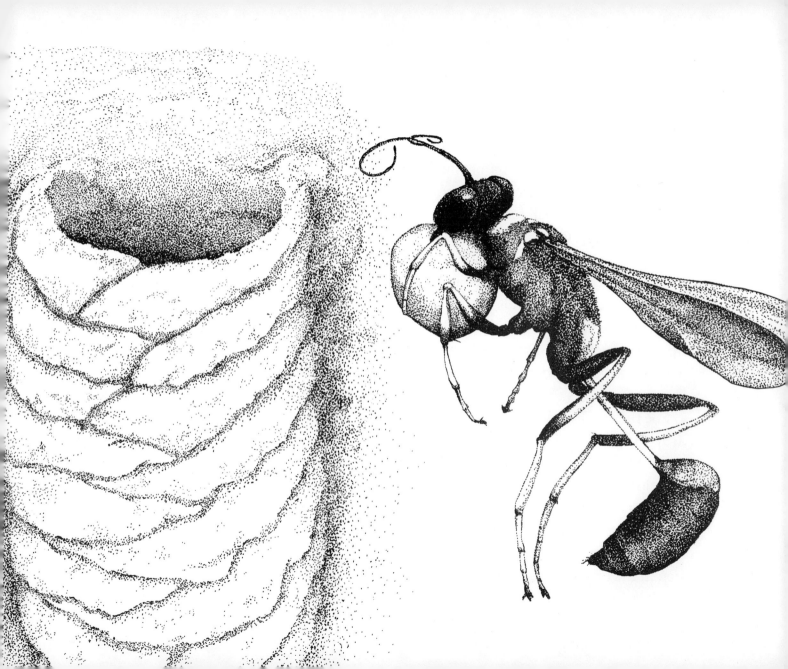

Working with her head and legs, she shapes each ball into a short arch.

The wasp makes many trips. She may gather thirty or forty balls of mud. She sticks each arch into place so that together they form a strong wall for the cell.

But the wasp is not yet ready to lay her egg. First, she must supply food for the young wasp, or grub, that will hatch.

She hunts until she finds a spider. Then she attacks it with the sharp stinger on the tip of her abdomen. She stings the spider with great skill, so that she doesn't kill it. She simply wants to paralyze it.

She flies to the cell with a live spider and drops it inside. Then she finds several more. She lays an egg on one of the spiders and closes the cell with a lid made of mud.

Insects act by instinct. They cannot go to the next step without finishing the one before. They cannot leave out, or repeat, a step in the process of breeding their young.

The mud dauber in the picture has completed three cells side by side. She fills each with spiders and lays a single egg.

The picture shows what happens after the cell is sealed. The first cell has not yet been filled with spiders. The second shows the single egg with the spiders. In the third cell the grub has hatched and finds itself surrounded by fresh food. It eats the spiders and grows rapidly.

If the mud dauber had killed the spiders before stocking the cell with them, their bodies might have rotted before the young wasp hatched and had time to eat them all.

In two or three days the grub turns into a pupa. In this state, its body goes through many changes until it takes the form of an adult wasp. Now it is ready to push off the lid of the cell and fly away.

Soon another female wasp will be looking for muddy places so it can build a cell of its own.

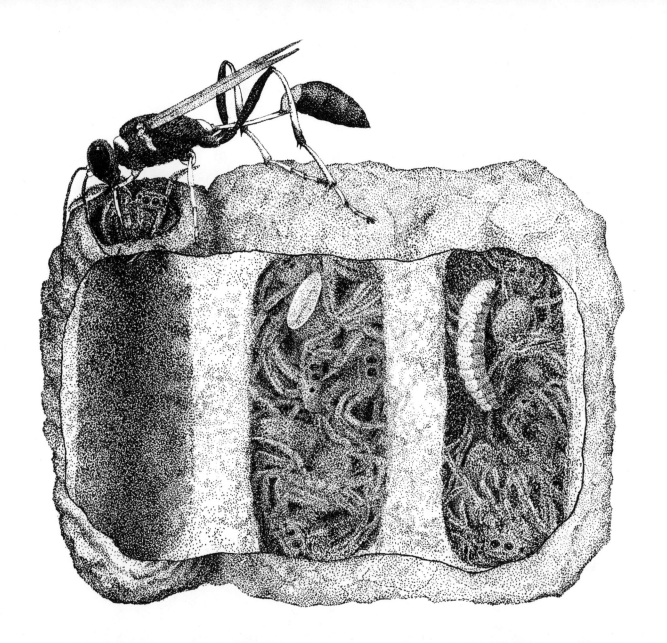

FIELD CRICKET

Our senses give us information about the world. Human beings carry their organs of sight, smell, taste, and hearing in their heads. But an insect's head is too small to have room for all of these complicated sense organs. The field cricket, for instance, listens with its knees.

This is very convenient because a cricket's body, like those of other insects, has many tiny breathing pores in its outer covering, or "skin." Fresh air is taken in through the pores and carried throughout the body by a network of tubes.

Some of these tubes widen into air pockets. A field cricket has air pockets near its knees. Sound vibrations from outside

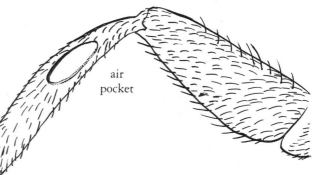
air pocket

are passed through a special membrane at these pockets. The membrane works like our eardrums, sending vibrations to body cells that act on the information received.

The ears on a field cricket's knees are

very sensitive. They detect high-pitched sounds we don't even notice.

When a field cricket hatches from its egg, it looks much like its parents, except that it is smaller and wingless. It has no need to fly. Like its relatives, the grasshoppers, it has muscular legs. A field cricket is about a half-inch long when grown, but it can jump two feet.

By this time it needs wings, though it seldom flies. A field cricket uses its wings for singing.

Insects do not have voices as we do. They communicate by scraping parts of their hard bodies together. Male crickets sing by rubbing their wings against each other.

Their front wings have special tools for

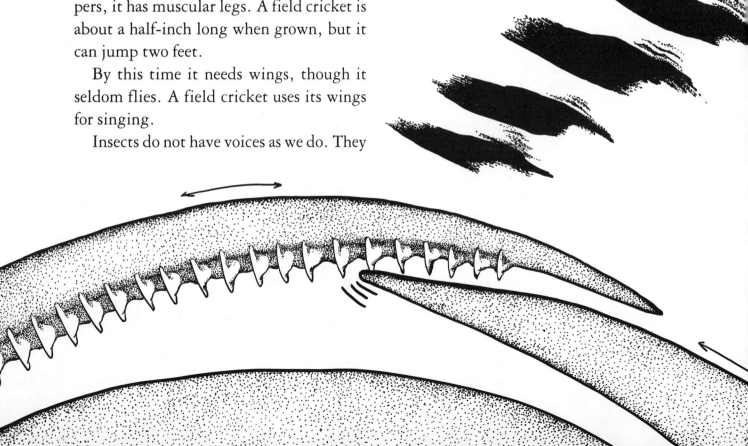

making loud sounds. Each has a flat, rough spot and a vein with a sharp, file-like edge. The male rubs one wing over the other so that the toothed vein scrapes the rough spot and produces a song.

The wings vibrate nearly five thousand times a second. The field crickets and their relatives never seem to tire. One relative, the katydid, may rub its wings together fifty million times during the summer.

"Treet-treet-treet," the field cricket seems to sing. But scientists, who have recorded the song on sensitive instruments, say we cannot detect all of the complicated song the cricket hears.

When the male begins to sing, he takes the first step in the mating ceremony. His song warns other males to stay away. It also attracts a female.

The ears on the female's knees pick up slight differences in sound. She will not respond to the song of any other kind of cricket. She waits to hear exactly the same song that male field crickets have sung for millions of years.

Without that sound, none of the other actions that come after it will take place. The stimulus—the important link—is missing. When she finally hears it, she mates and lays her eggs in the ground.

BURYING BEETLE

Small animals live around us in great numbers, but most of them live in a secret way and we seldom notice them. These small mammals, birds, and reptiles have short lives and many of them die every day from accidents and disease. What happens to their bodies?

The natural world has its own way of cleaning up. We see ants attacking a small dead creature, or crows feeding on a dead squirrel along the highway. But perhaps the most efficient gravedigger is the burying beetle.

The burying beetle is always searching for dead animals. This one has found a dead vole. Soon it is joined by another beetle that will become its mate.

The burying beetle is a dark-colored insect, about an inch long, with bright red

markings. There is a tuft at the end of its long antennae. Like many other insects, it detects odors with its antennae and other sensitive parts of its body.

As we saw earlier, the mud dauber wasp looks only for certain kinds of tiny animals such as jumping spiders that she can carry to her cell. The burying beetle searches its world for almost any kind of small animal. Mice, birds, lizards, snakes, moles, frogs, shrews—the burying beetle smells their rotting flesh and finds them wherever they have died. Like the mud dauber, it must provide food for its young.

The dead vole is many times larger than the beetles. It is too heavy for the insects to carry away. So these burying beetles go right to work and bury the vole's carcass on the spot.

They are not graceful insects. Their bodies are stout. Strong muscles, attached to hard shells, give their movements a strength and heaviness that is surprising in such small creatures. But they work rapidly.

The beetles force their way beneath the vole's body and begin digging. They work furiously, bringing all six legs into action. They drive their front legs into the hard earth like shovels. With the other four legs they sweep aside the loosened dirt. Their sharp mandibles, or jaws, cut through grass and slender roots.

If a person came along at that time, the scene would appear very mysterious. The eye would see only a dead vole. But the vole would be moving! The carcass would twitch. It would rock from side to side. It would be slowly sinking into the earth.

The burying beetles are out of sight, underground. They tug at the carcass and dig away the earth beneath it. The loose dirt is pushed up toward the surface. The carcass slowly sinks. As the carcass sinks, loose dirt piles up around it.

The job takes only a few hours. The burying beetles work tirelessly. The carcass disappears as the pile of loose dirt finally slips into the hole over it. Only a low mound of dirt shows where the carcass used to lie. The person watching from above might think that the vole had buried itself.

But the beetles are still hard at work underground. They go on digging away the earth beneath the carcass, letting it sink deeper. With their mandibles they strip away its fur.

The female lays her eggs in a tunnel that she digs near the carcass. The grubs, when they hatch, will find the food the parents prepared for them before flying away.

The burying beetles' powerful legs, mandibles, and backs allow them to complete their work. Dig they must, their instinct tells them. By following this instinct, they make our environment a cleaner, healthier place.

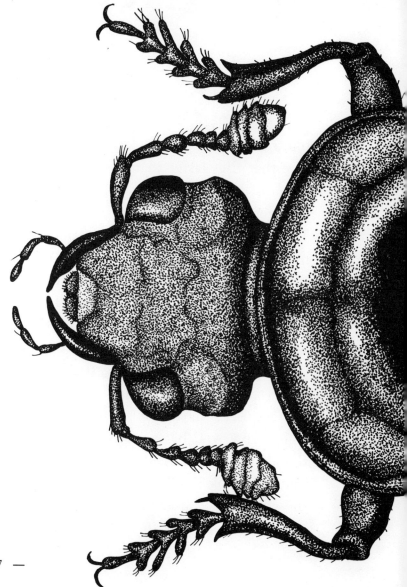

DRAGONFLY

The dragonfly is well named. It behaves like a winged dragon—a large, fierce creature that flies faster than any other common insect. Darting from its perch, it flies out over the pond to capture other insects. The dragonfly pictured here has caught a mosquito.

These insects are so ancient that they have not developed along the lines of their modern relatives. Most insects fold their wings when they land. The dragonfly clings to a leaf with its long transparent wings spread out in the sun.

Few insects have such beautiful wings. They grow well back on the dragonfly's thorax, and are crisscrossed with veins so that they look like little nets. Even when the wings are tinted with soft colors, we can see right through them.

Yet these beautiful wings become perfect flying machines when the dragonfly springs into the air. Its thorax is packed with powerful muscles for flight. The veins brace the wings so that they can vibrate about forty times a second.

Like most other adult insects, a dragonfly has two pairs of wings. But it has not developed the little hooks that enable more modern insects to link their wings and flap them together. A dragonfly flaps the two pairs separately, one pair rising while the other beats downward.

Even so, a dragonfly is a remarkable flyer. Scientists have timed its flight at sixty miles per hour. It soars like a gull, dives like a hawk, and even flies backward.

The dragonfly spends most of its life in flight. It is a great hunter, busily flying out

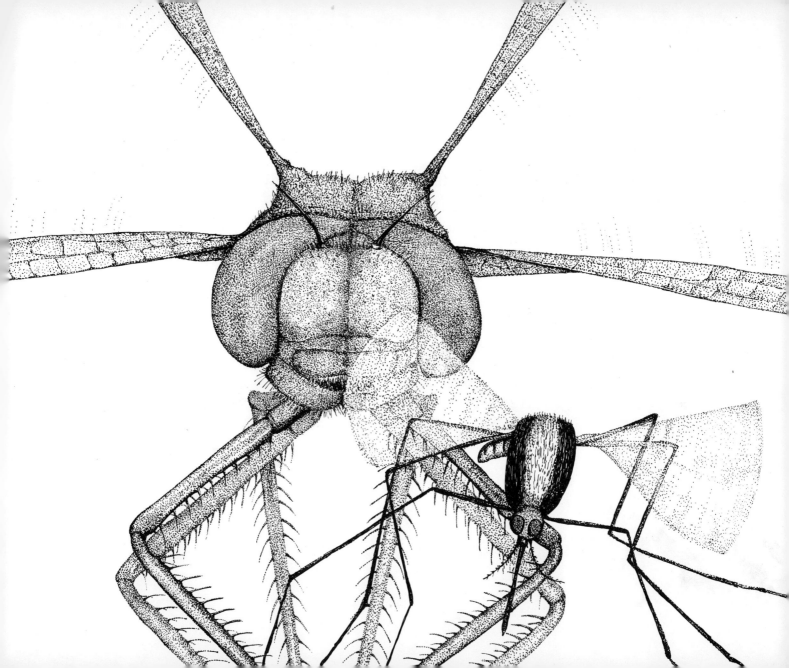

over the lakes and ponds where it is usually seen. It hunts over a familiar route, searching for food and driving away other males of its own kind.

This area is the dragonfly's territory. He mates with females that enter there, but fights the other males so fiercely that sometimes his legs and wings are damaged.

No small insect is safe from this hunter.

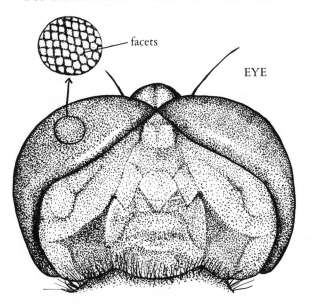

It eats so many mosquitoes that it is sometimes called a "mosquito hawk." But it flies out and attacks larger creatures such as moths, butterflies, bees, and horseflies.

A dragonfly is made for hunting on the wing. Aside from its wings, its eyes are its most prominent part. The antennae are small because it does not depend on them as much as most insects do.

The eyes are enormous, covering much of the head. Like other insects, the dragonfly cannot close them, even when at rest. They are called compound eyes because they are made up of many tiny surfaces called facets.

Some kinds of dragonflies have eyes with as many as thirty thousand facets. Each of these mini-eyes points in a slightly different direction, enabling a dragonfly to look in almost every direction at once. Its brain takes all these little pieces of sight, and builds them into a single picture of what is going on in the immediate vicinity.

Insects are usually very nearsighted. They can only make out images a few inches away. But a dragonfly can see movement more than forty feet away. This keen sight helps it to capture other fast-flying insects, as well as to escape from its own enemies such as birds and bats.

The dragonfly's feet are designed for hunting, too. They are placed very far forward on the thorax, right under the head. They are not much use for walking or running, but such a skillful flyer does not need them for that purpose.

The legs have a more important use. As the dragonfly hunts over the pond, it bunches its legs so that they form a little basket. When it sees a small insect, it darts forward and sweeps it out of the air. Then, with its front legs, it brings the insect close to its strong jaws and eats it.

Sometimes, on calm days, we see a dragonfly skimming low over the water and dipping its tail beneath the surface. It is a female dragonfly, laying its eggs.

The eggs drop to the bottom of the pond. The nymph, or young dragonfly, lives underwater. Then, one day, it will crawl up a blade of grass and dry its new wings in the sun. Another hunter is about to fly out over the pond.

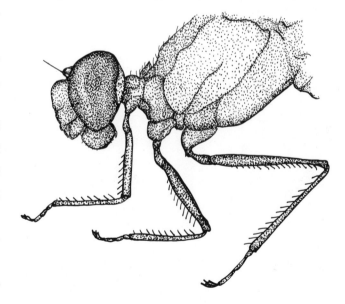

APHIDS

Aphids are weak, almost defenseless, creatures that would seem to stand litle chance in the fierce jungle of insect life.

An aphid is only one-quarter of an inch long, with slender antennae on its small head and three pairs of feeble legs on its short thorax. Its abdomen, however, is large and plump. At its tail is an organ that pumps out the sweet juices for which it is famous. Above this organ are two knobs that form a strange defensive weapon, as we shall see.

Perhaps millions of years ago the aphids' ancestors had strong mandibles to chew their food. But now these jaws are joined into a long, tube-like organ on the front of

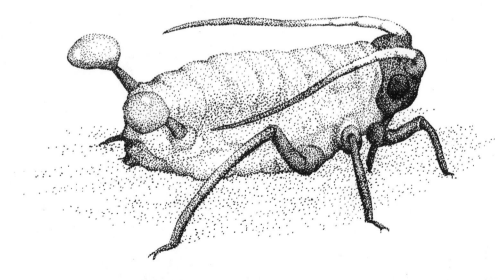

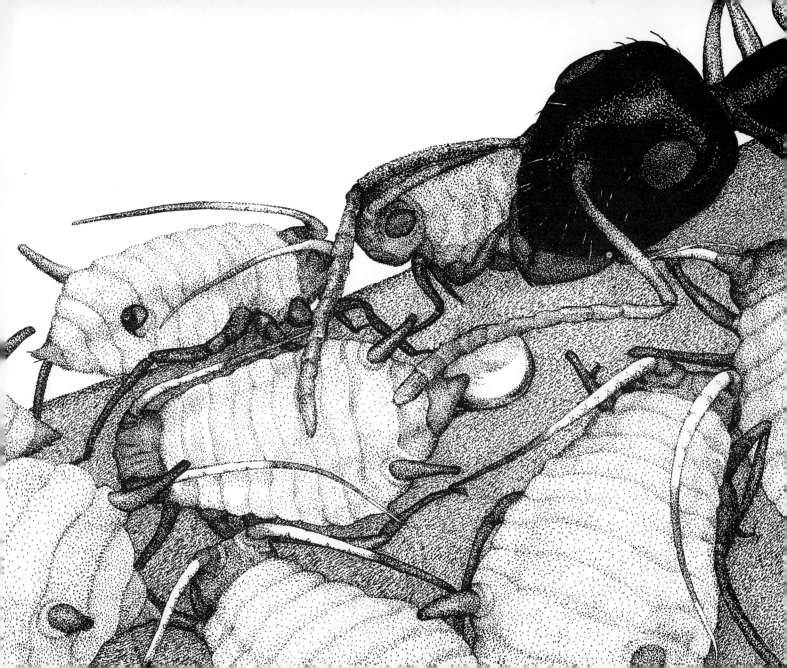

the head called a proboscis. Instead of chewing its food, an aphid uses its proboscis to suck juices from plants.

In the picture opposite, the aphid has unrolled its proboscis and jabbed it into a green leaf. The proboscis sends out an extremely thin tube that snakes through the leaf's cells, which are seen here greatly enlarged. An aphid passes the tube around some of the cells until it finds the ones it prefers. Then it sucks up the juices.

These juices are digested by the saliva in another tube. Finally the aphid squirts waste from this food through the organ at its tail. This sweet liquid is honeydew.

Several kinds of ants love honeydew. They look after large numbers of aphids. They often prod the aphids with their antennae, urging them to feed more rapidly. The ants then gather the honeydew.

Ants even move aphids to other plants. These small green insects increase so rapidly that a plant they feed on soon begins to die. As a rule, aphids are wingless. If there are no ants to help them move, what can they do?

A marvelous thing happens. Some of the female aphids grow wings. They are able to fly off to another plant. There they start a new colony.

In return for honeydew, ants protect aphids from their enemies. Ladybird beetles eat many aphids. But some aphids can help themselves. They point the knobs on their abdomens at an enemy and squirt a waxy liquid into its mouth. This liquid clogs the enemy's mouth so it cannot eat the aphids!

Scientists have watched an ant help an aphid pull its long proboscis out of a plant when enemies came near. The aphid was too busy doing what it is supposed to do, sucking plant juices and making honeydew, to run away.

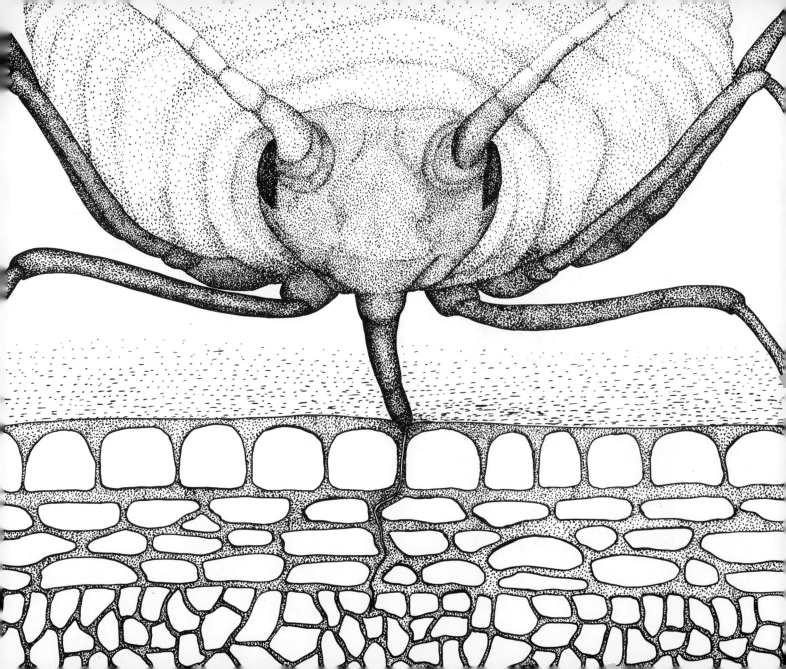

FIREFLY

The sun sets and darkness begins to fall over the meadow. A female beetle crouches motionless, clinging to a blade of grass.

Suddenly, a greenish-blue light flashes in the trees above the meadow. It flashes again. Other lights flash on and off, rising

in the night air. The tip of the beetle's abdomen glows as she flashes a reply.

These beetles are fireflies. The distant lights over the meadow are made by male fireflies searching for a mate. When the female recognizes the familiar signal, she responds and attracts a mate to her side.

The brown and yellow fireflies in this meadow are three-quarters of an inch long. They are just one of a hundred kinds of fireflies that live in the United States.

Each kind has a secret code of its own. It flies higher or lower than other fireflies. Its flashes may be longer or shorter than the others', and so are the pauses between the flashes.

This female, waiting for a mate, replies only when she sees a male flash rapidly four

or five times in a row. She always waits for several seconds. Then she flashes her reply two or three times.

A firefly carries its lighting mechanism at the tip of its abdomen. This light, unlike the ones that we manufacture, gives off no heat.

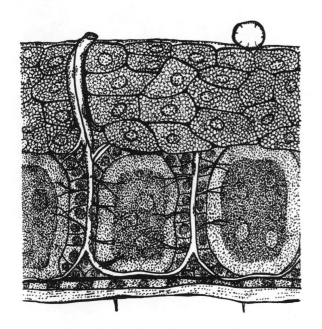

There are two curious drawings on this page. The one on the left shows a part of the firefly's abdomen as you would see it if you looked at it through a microscope.

Air is carried by the tubes into the large cells in the bottom layer. It mixes with chemicals to produce a light.

A firefly controls its flash of light by controlling the supply of air through the tubes. The flash is bright or dim, depending on the amount of air that reaches the cells.

The other drawing shows the pattern of signals flashed by different kinds of fireflies. Each kind signals for a mate with a code of its own. These are some of the signals recorded by a scientist. The second row shows a series of squiggles. They are the signals flashed by a firefly while it was shaking its abdomen!

The male fireflies are advertising for mates. Some kinds of fireflies send a series of signals, anywhere from two to eleven flashes. Others send only one flash at a time. A female tells if this is a signal from a male of her own kind by the length of time between single flashes.

Only the males fly when they signal. Rising on their fragile wings, they flash their lights constantly, trying to attract a mate. The female, which is flightless, is often called a glowworm. After she attracts a mate, she will lay her eggs on the ground.

In meadows everywhere, fireflies busily flash their lights, sending out their signals in code.

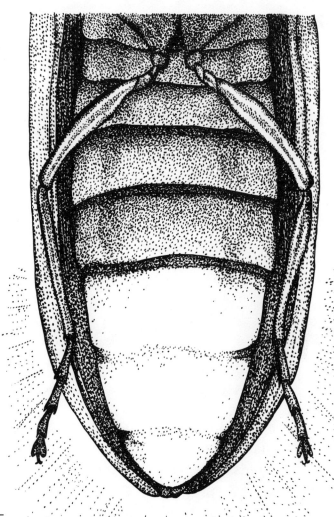

HONEYPOT ANT

There are more than two thousand kinds of ants. Almost any small plot of ground on earth holds thousands of individual ants, all scurrying around, engaged in different kinds of work.

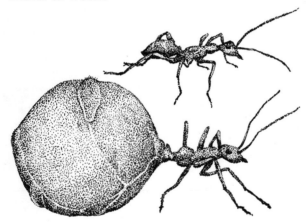

Ants hunt other small creatures, gather bits of wood or fruit, carry away seeds or shreds of leaves, and tend aphids. But strangest of all these "workers" are the honeypot ants which spend their lives serving as tiny barrels of honey.

Like other kinds of ants, honeypots live in colonies. These colonies are made up of closely related individuals. Each one performs a special job. The queen of the colony does nothing but produce eggs. Other ants feed the queen, or guard the nest, or hunt for food.

Honeypot ants live only in dry areas of North America between Colorado and Mexico City. Their nest is always underground. They collect honeydew from aphids and from plants where gall wasps lay their eggs.

This sweet liquid is usually available for only two or three months a year. The ants need to store honeydew to prepare for dry periods. Bees store honey by making hon-

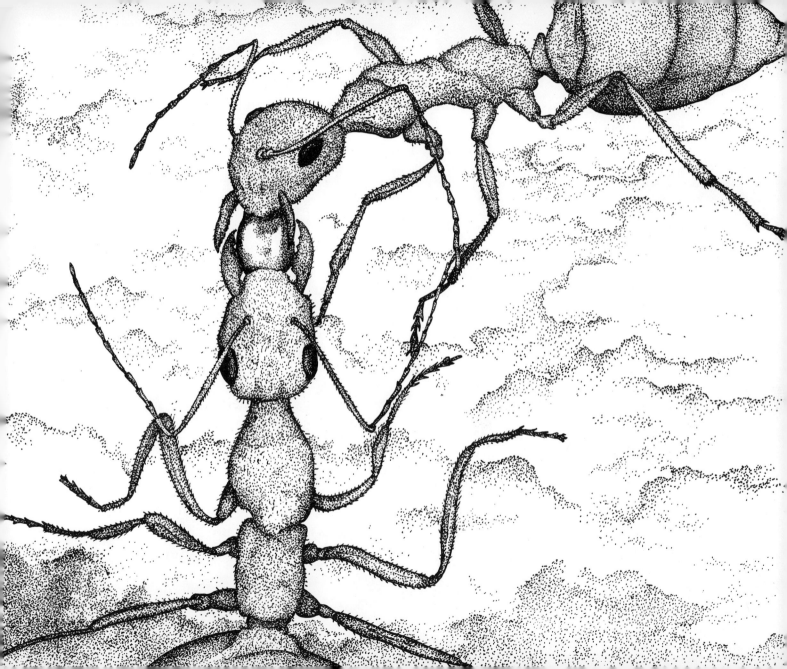

eycombs from wax. Honeypot ants, which cannot produce wax, have found another way to store it.

The colony's workers are females and look exactly alike at first. But only some of them go out to gather honeydew. When they return, they are met by other ants. From habit, they feed one of these workers.

Several other returning ants feed her, too. If she continues to accept honeydew, her future is settled. She becomes a honeypot.

The returning ants continue to feed her. Her body's covering stretches. Her abdomen grows larger and larger until it becomes the size of a pea.

After about a month, she drags her swol-

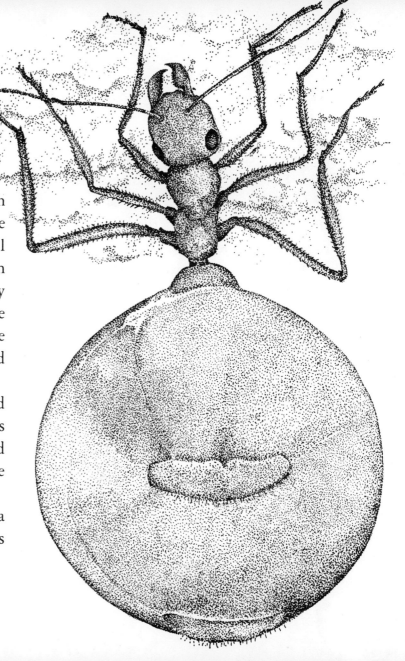

len body to a small chamber, about an inch high. As the pictures show, she has become a different-looking ant than the normal workers. She joins the other honeypots in the chamber, hanging from the ceiling by her feet. She will never move again. The segments of her abdomen, which were once close together, are now dark bands spread far apart.

The returning workers continue to feed the honeypots. Sometimes one of them falls to the ground. If she becomes too full and bursts, the other ants rush over to drink the honeydew.

The workers remain busy. They have a safe place to store their honeydew for times of drought.

HONEY BEE

A honey bee has landed on an apple blossom. In the distance is the hive, where the bee will fly with the nectar she gathers from the flower. This is the busiest period of the worker bee's life.

All the important work of the colony is done by females. The queen bee lays eggs furiously, often more than a thousand a day. A few eggs hatch into males, called drones, which mate with the queen. But most of the larvae are females.

At first, they are fed by older bees. They grow into workers. They spend more than half their lives in the hive, feeding new larvae, repairing the hive, and making honey.

Finally, these worker bees fly from the hive to gather nectar and pollen for food. Nectar is the sweet liquid found in flowers. Pollen grains form seeds when they reach the proper part of the flower. By helping spread pollen to other flowers, bees "pay" for the nectar and pollen they take.

The bee thrusts her tongue into a flower. She sucks nectar into a special stomach that holds the honey until she returns to the hive. She also collects pollen. She uses the spur on her middle leg to brush pollen grains from her body and pack them into the hairy baskets on her rear legs.

When she flies to another flower, some of the pollen that is stuck to her body rubs off and pollinates the plant. The plant can then produce its seeds.

The bee returns to the hive. She packs her pollen into storage cells, where it is mixed with honey to help feed the colony.

She also brings up the nectar from the honey stomach for other workers.

On this page you see a honey bee's head, greatly enlarged. In drawing (1) the bee is ready to feed from a flower. Its large eyes are big dark spots on each side of its head. The antennae droop in front of the eyes. The jaws, which move sideways, are open.

A sucking tongue is part of the long, dagger-like structure called the proboscis. When not in use, the proboscis can be drawn up under the chin (2). It moves like a hinge. Usually it is closed, but sometimes the parts open (3) and we can see the long hairy tongue.

The tongue is like a telescope. When the bee sucks nectar from a short flower, the tongue extends only a short distance (4). If nectar lies deeper, another part of the tongue slides out (5), making it longer.

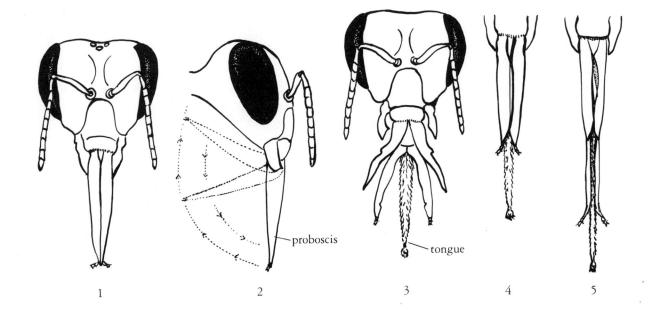

1   2   3   4   5

A busy bee needs special tools to take care of her chores. A honey bee's hairy legs do just that.

Close to the head is a first pair of legs (2). They are covered with stiff hairs for brushing the head and face.

These front legs have a small notch, and opposite it a comb. An antenna fits into the notch (1). The bee moves her leg to scrape the antenna clean against the comb.

The second pair of legs (3) has a spur, or comb, at about the same point as the first pair. A bee often swings these middle legs up over her back or down under her body to clean her wings and body.

The third pair of legs (4) is covered with stiff hairs. These hairs form the sides of the pollen baskets (5).

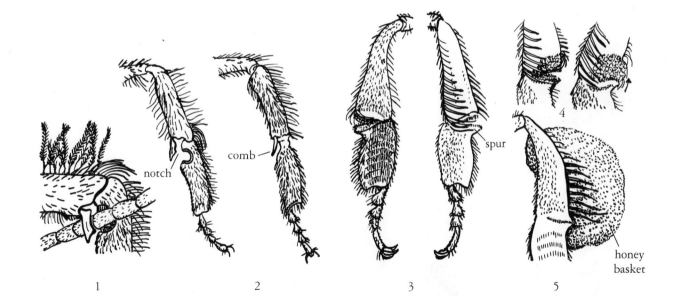

Honey bees buzz busily between the hive and a stand of dandelions. Pollen grains stick in the dense hairs that cover their bodies.

A bee gathers pollen as well as nectar. She flies back to the hive. The pollen will be fed to the bee larvae. The nectar from her honey stomach is given to other workers. They work the nectar with their own tongues to remove most of the water. It becomes the rich food we call honey. There will be plenty of honey for the colony.

If we consider its size, there is no animal —even a horse or an elephant—that has more strength or energy than an insect.

The amount of energy we use at rest is called our basal metabolic rate. Scientists say that a fine human athlete increases this rate twenty times for a short period when performing a sport. But a bee increases its basal rate over fifty times for many hours.

Like other insects, the honey bee has a

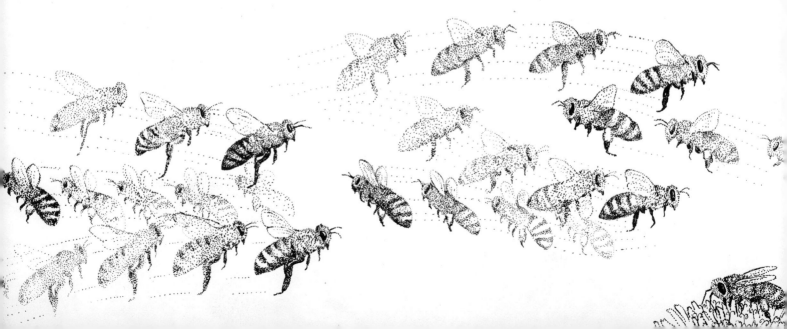

sturdy outer skeleton. Strong muscles, attached to her thorax, provide power for her wings.

A hive may have fifty thousand workers and need five hundred pounds of honey during the summer. The workers that gather honey seldom rest. They fly many miles each day between the hive and the meadow, wings beating 250 times a second.

Like many other insects, bees have two pairs of wings. The wings are locked together in flight with small hooks so that they beat as one pair and provide a broad wing surface. Each wing has a stiff frame, like that of a kite.

Even so, the wings take a terrible pounding. After about two weeks, their edges are ragged and broken. The bee, worn out by work, dies.

But with the help of her legs, wings, and tongue, the honey bee packs many lifetimes of work into her brief summer.

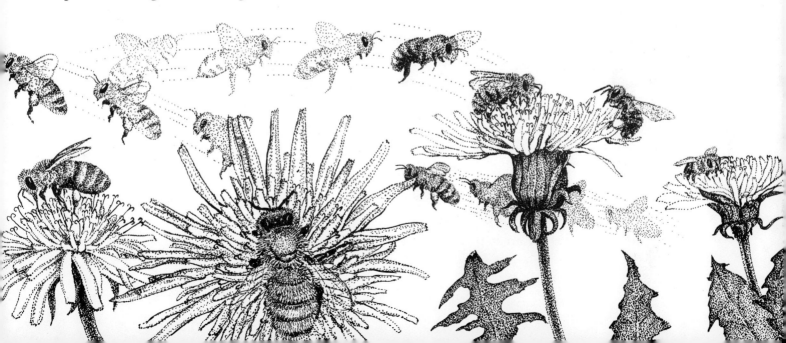

TENT CATERPILLAR

Scientists call young insects by different names. Young ants and beetles are known as grubs, and young flies as maggots. The young of moths and butterflies are called caterpillars.

We see insects often at this stage of their lives because they are so busy. Caterpillars are really eating machines. They spend their time chewing leaves and growing larger and fatter. Then they spin cocoons and change into moths or butterflies. This final stage of their lives is usually brief, just long enough to mate and lay eggs.

The tent caterpillars in the picture here hatched from eggs that were laid on a cherry tree by a moth. Their soft, round bodies are densely covered with hairs. They have tough jaws for chewing leaves. They were only one-tenth of an inch long when they hatched, but they are growing rapidly.

Tent caterpillars, besides eating, spend much time spinning silk. When they are still young they spin a path from their home in the tent to the leaves they eat. The silk path makes it easier for them to cling to a branch, and then find their way home.

The caterpillar in the small picture on the next page is laying down silk. Below its mouth is a hollow tube called a spinneret. This is the tool with which it makes silk from a sticky liquid in its body.

As the liquid comes from the spinneret, it hardens and sticks to the branch. The caterpillar then bends its body to one side to draw out a long thread from the spinneret, as if it were unwinding thread from a spool. It crawls forward, attaches the sticky thread, and repeats the process.

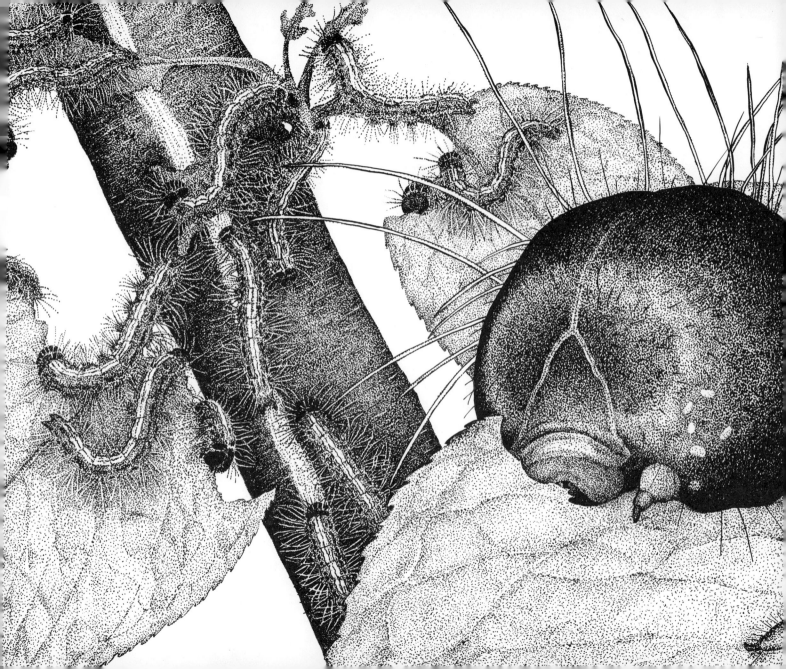

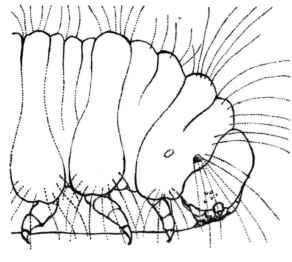

spinning silk

As tent caterpillars grow older, they begin to work together. They feed and spin on a regular schedule. They choose a place for their tent between two branches of a tree. The tent is small at first, but as these insects feed and grow, they enlarge it.

Each day is the same. The caterpillars leave the tent and meet on its roof. From there they crawl onto nearby branches and eat leaves. In the middle of the morning they return to their tent and work together, spinning new walls. Then they crawl inside to rest.

The afternoon is much the same. The caterpillars feed, spin, and rest. In the evening they all feed until late at night.

The picture opposite shows what a busy place the tent is during the day. Silk trails cover the branches. Caterpillars crawl over the tent, while the dark mass at the center shows where they rested inside.

As the caterpillars grow, they shed their old coverings. Inside, the tent becomes littered with these skins. Their lives as caterpillars are ending.

They leave the tent, one by one. Each caterpillar drops from the tree and finds a sheltered place to spin its cocoon. The cocoon seems as quiet as a grave. But inside, the caterpillar's body is busily at work, changing itself into a moth.

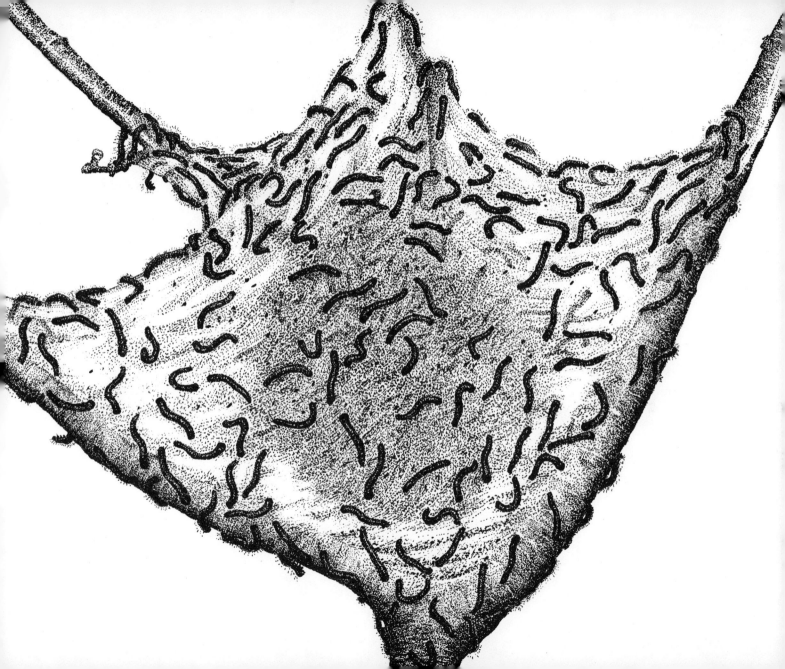

HOUSE FLY

Everyone hates house flies. We may watch in wonder as one of them walks straight up a window pane or strolls upside down across the ceiling. But usually all we want to do is get rid of them.

And with good reason. Most of the billion or so kinds of insects in the world never bother us. Members of the fly family, however, often are pests. Horse flies and black flies bite us. Gnats and mosquitoes are also members of the fly family, and we know how pesky they are.

House flies, which are active in our homes much of the time, cannot bite us. They have no jaws. But they have other habits that annoy us, and may even harm us. Because of their constant busyness and the way their bodies are made, they are able to spread many diseases.

This is curious, because anyone who has watched a house fly closely knows that it spends a great deal of time cleaning itself. It scrubs itself as dutifully as any house cat. The fly rubs its front legs together rapidly. Then it rubs them over the top of its head.

When it feels that its head is sufficiently clean, the fly extends its middle legs forward and gives them a good brushing with the front legs.

That done, it goes to work with its hind feet. It rubs them together and, when they are clean, uses them to brush its wings, thorax, and abdomen.

What a clean little creature, we might say to ourselves. But neither we nor the fly can see what it has missed. Sticking to its stiff body hairs and the pads of its feet, and living in its digestive system, are mil-

lions of nearly invisible bacteria and other germs.

Few other creatures are better equipped to pick up germs and carry them to where they may do the most harm. The house fly is a very active insect, with the habit of investigating almost everything in its surroundings. It has a stout body, with enormous eyes. They are compound eyes, broken up into thousands of mini-eyes like those of the dragonfly—which, by the way, is not a fly at all.

Flying from one place to another, the house fly works very hard. It beats its wings about 330 times a second. It seems to have no preference where it stops to visit on its constant round of investigations.

It will visit a manure pile or a garbage heap, and a few minutes later land inside our home. When it does, it brings with it all the harmful bacteria its feet and body picked up along the way.

Sometimes it seems that a house fly barely needs wings. Its remarkable legs take it nearly every place it wants to go. It has both claws and sticky pads on the tips of its legs.

When the fly is walking on the floor or a tabletop, it uses its claws to grip the surface. When it reaches a window pane or the ceiling, it walks on the pads of its feet.

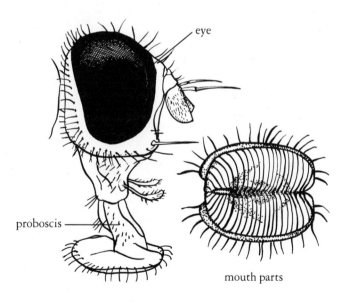

eye

proboscis

mouth parts

These pads are coated with a sticky substance. The fly takes a step, and then pulls its foot loose.

A house fly also tastes with its legs. If it walks over some kind of food, the taste organs in its legs tell it if it is good to eat.

Because a house fly has no jaws, it can take only liquid food. Yet it does not have to turn away from a bread crumb, a sugar cube, or a slice of steak. The house fly's special mouth parts are able to convert almost any food to a liquid.

These mouth parts are contained in its proboscis, or snout. When it is not eating, the proboscis is folded back under its head. There are two flaps at the tip of the proboscis and above them there is a long tube.

When the fly finds solid food, such as meat, it rasps or rubs it with the flaps on the proboscis to get at the juices it contains. The fly then sucks up the juices through the tube.

If the food is still not manageable, the fly brings up a drop of liquid from its body. This liquid usually dissolves the food and the fly sucks it into the tube.

House flies grow sluggish when cold weather comes. Most of them die. But they have hidden their eggs well, and like it or not, flies will be with us when spring returns.

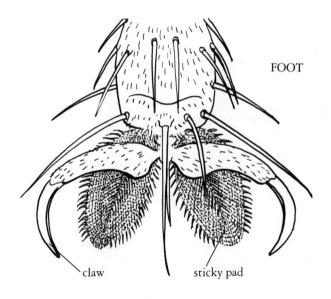

FOOT

claw   sticky pad

LEAF MINER

When you first look at this picture, you may think it illustrates a long, strange-looking animal living in a thick wall. But the animal is far stranger than that. The drawing, greatly enlarged, shows a caterpillar, hardly bigger then the head of a pin, living inside a leaf.

Leaf miners are common insects. We are usually aware of them, however, only when we see a leaf with strange markings on it. These markings are made by the tiny insects inside.

Miners live in the leaves of almost every kind of plant. Some even live in poison ivy. Leaves make a convenient home. They are not simply flat. They are composed of an upper and lower surface that protect the insect, and layers of juicy, nourishing plant cells in between.

Many kinds of tiny wasps, flies, beetles, and the caterpillars of moths are leaf miners. One thing they all have in common is complete metamorphosis. That means that they pass through a complicated series of changes during their lives—making a cocoon in which they change from a crawling, rapidly eating young insect into a winged adult that mates and lays eggs.

The leaf tells the whole story of metamorphosis. Each kind of insect makes a distinct pattern as it tunnels between the upper and lower "skins" of a leaf. Scientists can often tell the kind of insect inside by looking at these markings. They also find the insect's egg, cocoon, and old skins.

The miner in our picture lives in poplar leaves. The adult moth lays its egg inside the leaf. The caterpillar is so flat when it

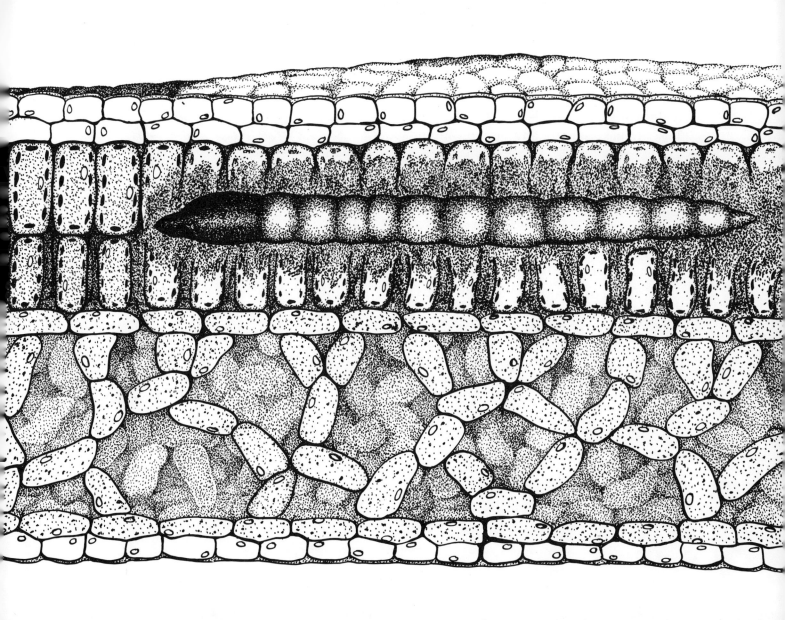

hatches, a scientist says, "It looks as if it was run over by a steam roller."

Like most leaf miners, it has no need for legs. It has scarcely any eyes or antennae, either. But it is well fitted for life in the leaf.

Its head is shaped like a sharp wedge, good for pushing the leaf apart as it wriggles through it. In the picture below, we see that its mandibles look like a circular saw. They cut through the plant's cells and suck their sweet sap.

The poplar leaf miner lives in the upper layer of cells. They are shaped like small columns. Other kinds of cells lie below, with many air spaces among them.

Most of the leaf soon turns shiny white.

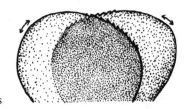

mandibles

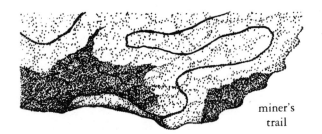

miner's trail

This is caused by the air that flows between its two surfaces as the miner pushes them apart. We can trace the miner's trail as it works its way through the leaf. The trail is made by the black frass, or waste, it leaves behind.

There is a great change in the caterpillar the last time it sheds its skin. The tunnel puffs out into a little pocket. The caterpillar's jaws shrivel up. It grows rounder and develops a spinneret.

The picture opposite shows how an edge of the leaf (far right) folds into a knot over the pocket. The caterpillar spins its cocoon and changes into a silvery moth, which chews its way out and flies away.

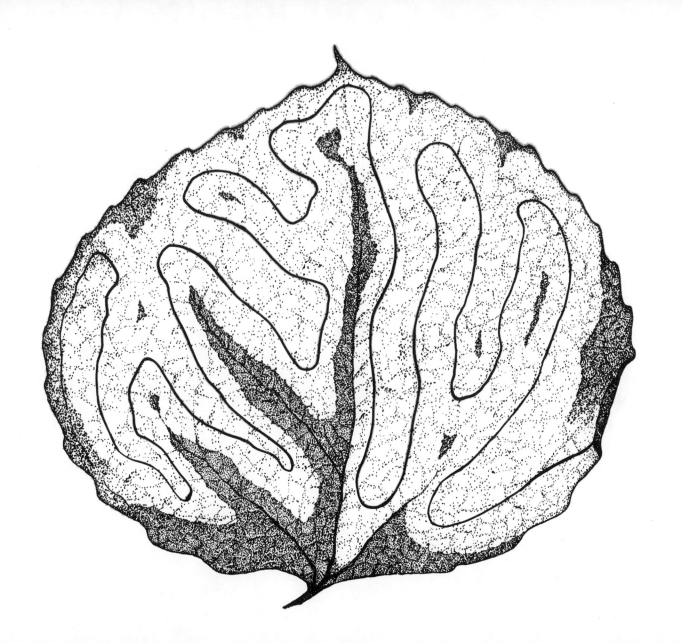

MONARCH BUTTERFLY

These pictures illustrate the life story of a monarch. This insect lives part of its life as a caterpillar, and part as a butterfly. Which is the true monarch?

The butterfly is the true monarch. The caterpillar is a creature that comes from the egg laid by the butterfly. It eats and eats. Finally it turns into a butterfly, mates, and lays eggs that hatch new caterpillars.

About nine out of every ten kinds of insects in the world go through this change that is called complete metamorphosis.

Insects benefit from complete metamorphosis. They are able to make the best use of food supplies. The caterpillar feeds on leaves. The butterfly feeds on the syrup in flowers called nectar.

The caterpillar, with its tough jaws and sturdy legs, is equipped for feeding. The butterfly, with its strong wings, is equipped to search for a mate and lay eggs.

The caterpillar is really two creatures in one. It contains two kinds of cells, which are the tiny building blocks that make up all living things. But only one set of these cells works while the monarch remains a caterpillar. As they grow larger, so does the caterpillar.

While the caterpillar eats and grows, the other cells play no part in its life. They are like buds on a plant that are waiting to open. When the caterpillar reaches its full size, it stops eating and turns into a pupa.

A pupa looks like a lump of dead matter. But inside, changes are taking place. Chemicals in the insect's body set these changes in motion. The old caterpillar cells die.

The other cells, quiet for so long, spring into life. They feed on the dead cells. They divide and form new shapes.

The stubby legs of the caterpillar become the long, slender legs of the butterfly. The strong jaws dissolve and form again as sucking mouth parts. Some of the new cells come together to build wings.

This is the life story that begins to unfold when the monarch butterfly glues her pale green egg to the underside of a milkweed leaf. All of the material to carry out these marvelous changes is enclosed in the tiny egg.

The butterfly has chosen the right moment to lay her eggs. It is early summer and the leaves of the milkweed plant are young and tender. Each baby caterpillar can easily chew the tender leaves.

As the caterpillar's cells grow, its body becomes plump. At each end of its body there are horns called whiplashes. Every few days it must stop feeding to shed the old skin that has become too tight for it.

The monarch caterpillar reaches its full size. A great change is about to take place. It spins a little tuft of silk and fastens itself to a leaf. It hangs upside down. Its head is curled up to one side so that it looks like the letter J.

The monarch caterpillar does not spin a cocoon as a moth caterpillar does. Its outer skin falls away. Beneath it is another skin that forms a case for the pupa, which in butterflies is called a chrysalis. Inside, great activity continues for about twelve days. The caterpillar disappears.

Now the case of the chrysalis cracks open. An insect with a completely new shape works its way out. The wings are small and wrinkled.

The monarch butterfly, which is not yet able to fly, clings to the broken chrysalis case. Soon fluids from its body flow into veins in the wings. The wings expand and harden.

The fluids flow back into the body and are replaced in the veins by air. The wings, covered by small round scales, are both light and sturdy.

Like many birds, some monarchs fly south for the winter. Until recent years no one was quite sure how far monarchs travel. Then scientists put paper tags on their wings. They gave each butterfly a special number so they could trace its flight. Some monarchs fly from Canada to Mexico on migration, nearly 2,000 miles!

Monarchs spend the winter sipping nectar from southern flowers. In spring the butterflies fly north again.

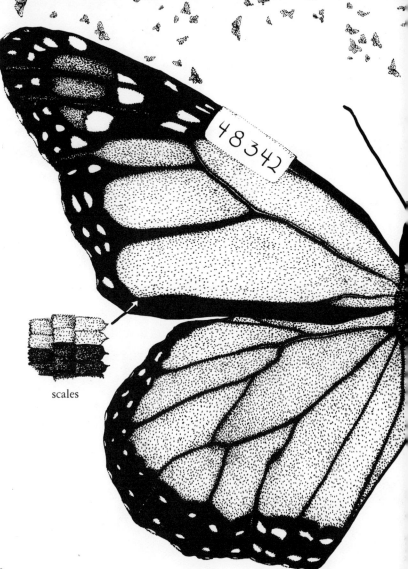

scales

ACORN WEEVIL

The acorn weevil, perched on the bumpy top of an acorn, seems like some creature from outer space. Why should any animal on earth develop such an outlandish shape?

The answer lies in the picture of the two acorns on this page. In the acorn on the left are three narrow tunnels that fan out into its center. At the end of each tunnel is a tiny white egg.

On the right is a picture of the grub that has hatched from one of the eggs inside the acorn. The acorn weevil's body is designed to get that grub into a safe place.

Insect eggs are everywhere. They are so numerous that we could never count even those in our own neighborhood. Yet they are so small, or they are so well hidden, that we are not likely to see them unless we hunt for them.

Many insects lay hundreds of eggs during their brief lives. Sometimes they lay them one at a time in different places. Sometimes they lay large numbers of eggs glued together on branches or the walls of buildings. There are round eggs and flat eggs, and eggs shaped like little barrels.

Female insects often lay eggs with the help of a narrow tube called an ovipositor. Sharp-pointed ovipositors are able to pierce leaves, and even trees, to lay eggs inside.

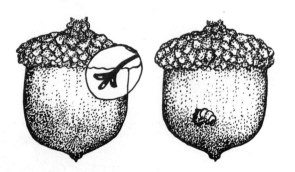

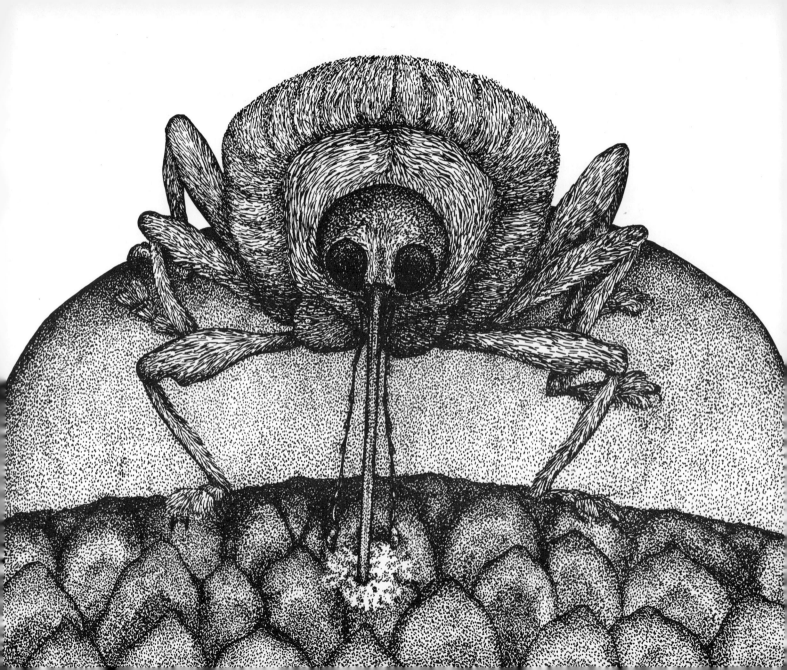

But the acorn weevil has found another way to lay its eggs in secret places.

No other family of animals in the world has so many members as the weevils. Scientists know the names of about 40,000 kinds of weevils. They say there are probably many more kinds that have not yet been discovered. The weevils belong to the group of insects called beetles.

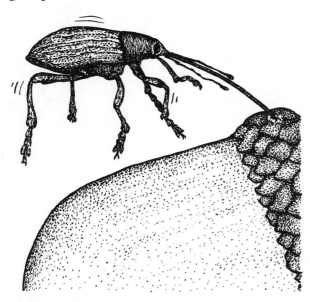

Weevils are specialists. That means that each kind of weevil prefers to feed on one type of plant. There are few kinds of plants that do not have a weevil that makes its living on it.

There are weevils that feed only on pine trees. There are weevils that feed only on cotton plants. And there are weevils that feed on acorns, the nuts of oak trees.

The acorn weevil's head narrows into a long point that is called a proboscis. The proboscis is nearly as long as the rest of the weevil's body. This insect's antennae, or feelers, grow out of the proboscis. If the antennae get in the way, the weevil can tuck them back into thin slots in the proboscis.

The acorn weevil's jaws lie at the tip of the proboscis. When the female weevil is ready to lay her eggs, she hunts for an oak tree with soft, young acorns. She pokes her proboscis here and there in the acorn's

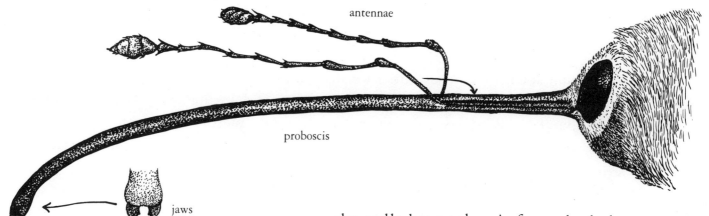

rough top, looking for a good place to cut into it.

The acorn weevil's jaws are strong and sharp. She uses them to drill a hole in the acorn. She works furiously, drilling into the soft inner pulp of the nut.

Once in a while a weevil's feet slip and she hangs by her proboscis. She will die if she cannot free herself.

The weevil drills a short tunnel that curves away from the main drilling. Then she pulls her proboscis from the hole. Turning around, she backs up to the hole and lays an egg into it.

When the acorn weevil's grub hatches from the egg, it will find itself safe from enemies, surrounded by nourishing food deep inside the acorn.

In autumn, the acorn falls from the tree. The grub cuts its way out and burrows into the earth. When summer returns, it becomes a pupa. It then emerges as an adult and, like weevils all over the world, it will mate and look for a place to lay its eggs.

CADDISFLY

The young caddisfly hatches into a world that is always moving past it. Gills, something like those of fish, permit this insect to breathe underwater in fast-running streams.

Most creatures that live in water are constantly moving, too. Fish, water beetles, and pollywogs swim through their watery homes, hunting for food. But the caddisfly in these pictures lives more like a spider. It weaves a net and waits for the stream to bring food to where it lives.

All of us admire the skill a spider shows in spinning its web. Hardly anyone knows

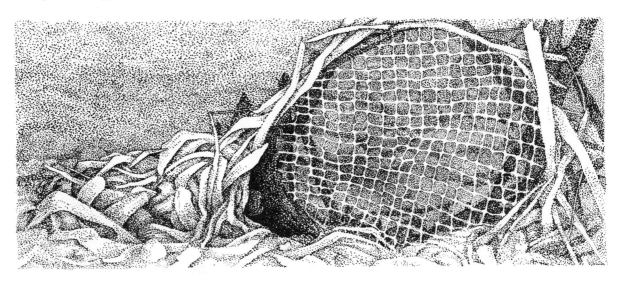

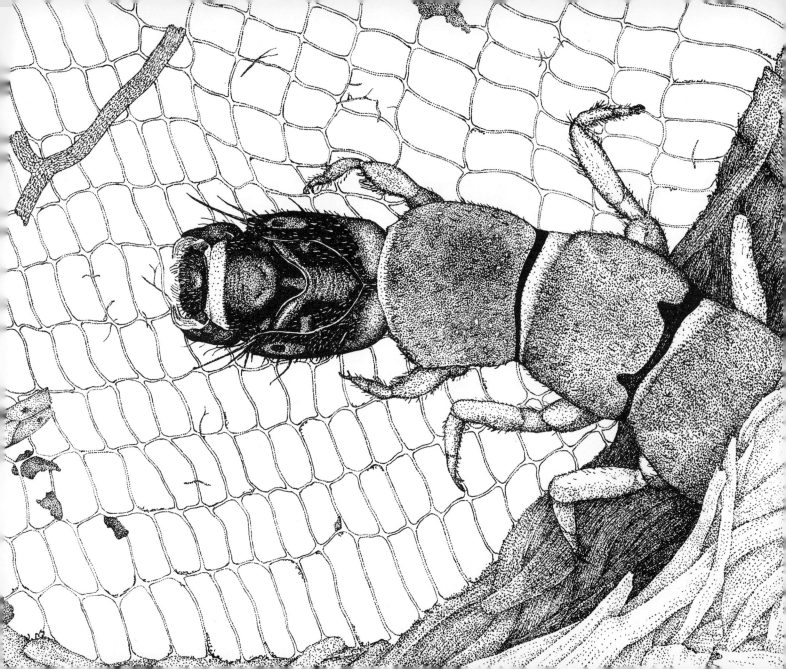

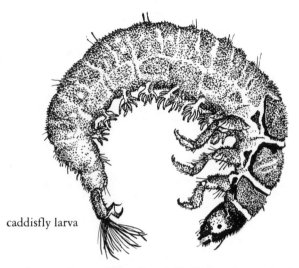
caddisfly larva

about the webs of caddisflies, though they are just as wonderful. This insect must work in the midst of rushing water and spin a net tough enough so that it won't be carried away.

Caddisflies vary in size from no bigger than the head of a pin to about an inch long. Only certain kinds build nets, and the nets are of different sizes, too. The largest are about half an inch across, though some are much smaller.

Before starting to spin its net, a caddisfly builds a shelter for itself. It collects small bits of material—fragments of leaves, sticks, sand, and shells—from the bottom of the stream. It binds them together into a little shelter near the place where it spins its net. It is a good hiding place because it looks like a pile of junk.

The picture on this page shows the underside of a larval caddisfly's head. The spinneret is located under its jaws. When it begins to spin its net, it works rapidly. It moves its head from side to side, drawing out the silk.

First, it fastens an end of the net to a log or a rock. The size of each mesh, or opening, in the net depends on the speed. In slow-moving streams, it spins small meshes. In fast streams, the openings are larger so the water will flow through easily and not damage the net.

A caddisfly spins its net in less than ten

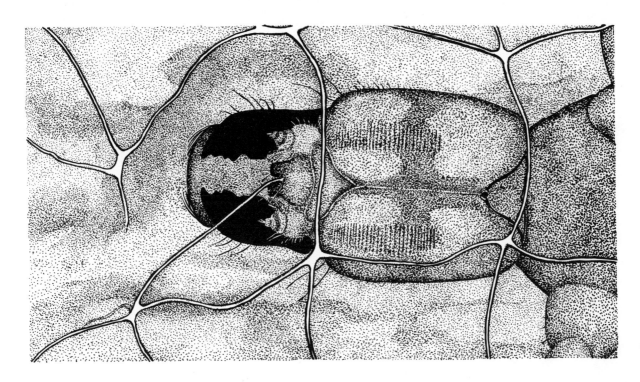

minutes. Tiny animals and plants are carried into the net by the flowing water. The caddisfly comes out of its retreat to feed. It has thick brushes around its jaws. It brushes the food out of the net into its mouth.

We think of winds and waves as great forces that change the earth. But caddisflies, like the other tiny creatures we have read about, are a force, too. Billions of insects, busily at work, quietly change the world around us.

SCIENTIFIC NAMES

Mud Dauber Wasp — *Sceliphran caementarium*

Field Cricket — *Acheta assimilis*

Burying Beetle — *Nicrophorus marginatus*

Dragonfly — *Libellula saturata*

Aphids — *Aphididae*

Firefly — *Photuris pennsylvanica*

Honeypot Ant — *Myrmecocystus mexicanus*

Honey Bee — *Apis mellifera*

Tent Caterpillar — *Malacosoma americana*

House Fly — *Musca domestica*

Leaf Miner — *Phyllocnistis populiella*

Monarch Butterfly — *Danaus plexippus*

Acorn Weevil — *Curculio rectus*

Caddisfly — *Hydropsyche morosa*